YOU CHOOSE

MACHINES

SONYA NEWLAND

WAYLAND

First published in Great Britain in 2024 by Wayland

Produced for Wayland by
White-Thomson Publishing Ltd
www.wtpub.co.uk

Editor: Sonya Newland
Series Designer: Rocket Design (East Anglia) Ltd

HB ISBN: 978 1 5263 2434 4
PB ISBN: 978 1 5263 2435 1

MIX
Paper from responsible sources
FSC® C104740

Wayland
An imprint of
Hachette Children's Group
Part of Hodder & Stoughton
Carmelite House
50 Victoria Embankment
London EC4Y 0DZ
An Hachette UK Company
www.hachettechildrens.co.uk

Printed in China

Picture acknowledgements:
Shutterstock: Agor2012 3, 25, d3verro 4, Catalyst Labs 5, Karinka 6, NoraVector 7, D things 8, lineartestpilot 9, bioraven 10, Freelancer Designer 11, hudhud94 12, GraphicsRF.com 13, Memo Angeles 14, Alphavector 15, Simple Line 16, Vectorium 17, johavel 18, The Toon Company 19, MaryValery 20, Basjan Bernard 21, Jemastock 22, Alexander_P 23, Susanto Anto 24l, rudall30 24r, ElsayedKaram 26, kosmofish 27, syawalan 28, Sylverarts Vectors 29, Studio Barcelona 30, igor malovic 31, Olga Kuevda 32, Kazakova Maryia 33, bsd studio 34, olegtoka 35, Redshinestudio 36, Vector Tradition 37, 63, kichikimi 38, ArtMari 39, 40b, 54, MoreVector 40t, Mechanik 41, Oleksandr Panasovskyi 42, angelh 43, brightshine_studio 44, Storyet 45, Azat Valeev 46, wenchiawang 47, Kate Shannon 48, Kailesta 49, Seita 50, Rvector 51, doddis77 52, vanillamilk 53, amazing studio Yogyakarta 55, Visual Generation 56, Hamdesain 57, Kyle Cr8on 58, BadBrother 59, zigzag design 60, Rainbow-Pic 61, siridhata 62, Ron Leishman 64, Anton Brand 65, Blue Planet Earth 66, mijatmijatovic 67, SkyPics Studio 68, SkyPics Studio 69, N.Petrosyan 70, GraphicsRF.com 71, Far700 72, Dzm1try 73, Icon Craft Studio 74, Dusan Pavlic 75, gerilya 76t, Aluna1 76b, art3 77, Magic Pen 78, Sudowoodo 79, rogistok 80, AVA Bitter 81, Visual Generation 82, puruan 83, Moriz 84, John T Takai 85, 88, nikiteev_konstantin 86, Tatiana Lazinina 87, Anastasiia Sorokina 89, drawww.studio 90, Leremy 91.

All design elements from Shutterstock.

In this book, you get to choose between two options about machines (neither of them usually very nice!). Study the pros and cons before making up your mind. Along the way, you'll discover some bizarre, hilarious and disgusting facts about all sorts of machines.

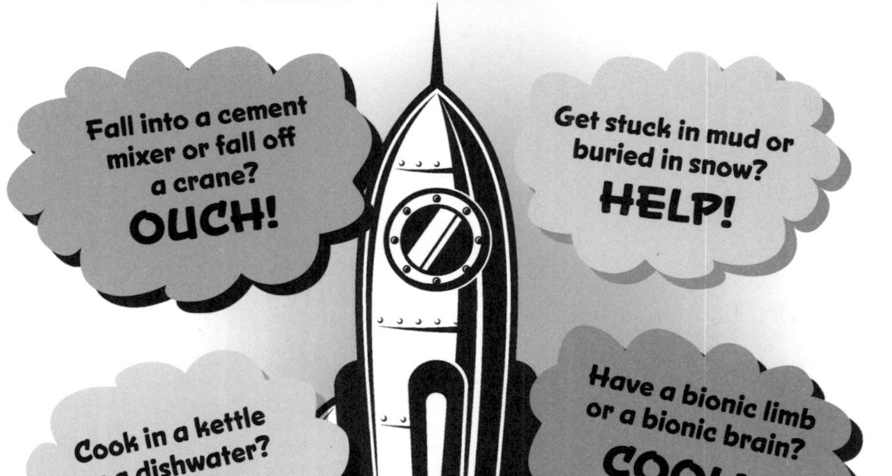

You choose!

Would you rather live on the International Space Station ...

Prepare to have your space invaded!

Living in space sounds exciting, but in reality it might not be much fun. To start with, it's pretty cramped on the International Space Station! For personal space, astronauts get a tiny cabin with a sleeping bag clipped to the wall.

The ISS orbits on a fixed path, so there's no way of changing course. Looking out of the window, astronauts see the same view over and over and over again. Plus, a new day dawns every hour and half up there, which would take some getting used to!

... or on a submarine?

Get a sinking feeling

On a submarine, you could travel to different places with the help of on-board navigational computers, but you couldn't enjoy the scenery. There are no portholes, so submariners have no view at all!

Submarines are amazing machines that can generate their own oxygen. In theory, they could stay underwater for years at a time, but really there's only room to store enough food for the crew for about three months. That's still a long time to stay submerged in close quarters!

YOU CHOOSE

See a new dawn every 90 minutes or no daylight for 90 days?

Would you rather get stuck at the top of a Ferris wheel ...

Fun fairground machine!

Ferris wheels are made up of a huge metal wheel that moves around a central point. People sit in seats attached to the rim of the wheel and enjoy a leisurely ride, round and round.

At the end of the ride, the wheel stops to let passengers off at the bottom. But that means that for a few seconds, some people are left swinging around at the top, 65 metres in the air. You can't help wondering – what if I get stuck here?!

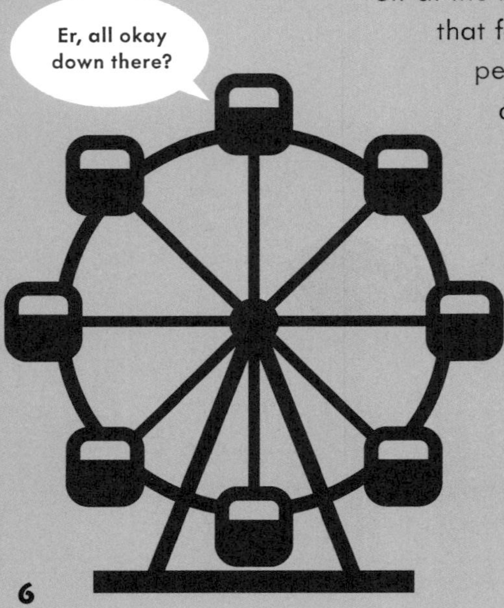

... or halfway down a zip wire?

Zip from place to place

Zip wires are simple machines. A cable runs between a higher point and a lower one. Strap on a harness, clip yourself to the cable – and off you go! Gravity pulls you down the line while a pulley reduces the friction so you pick up speed.

Basic physics says that you'll go faster the heavier you are, the steeper the slope and the longer the line. The faster the better really, as you wouldn't want to get stuck along the way – most zip wires are pretty high off the ground!

YOU CHOOSE

Stuck swinging in a seat or left dangling from a wire?

Would you rather only study the world through a microscope ...

Explore Earth's inner workings

A microscope is a machine for looking at small things. It uses lenses and light to make objects look larger, and allows us to see things that are invisible to the naked eye.

Without microscopes, we wouldn't understand how the cells in our body work, how diseases spread, how plants breathe or many other natural processes that take place all over our planet. Who knows what life-saving science microscopes might reveal in the future?

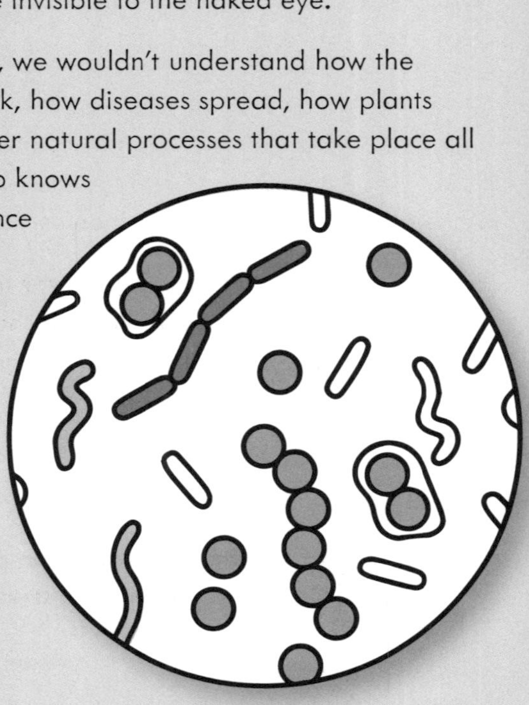

... or through a telescope?

Understand the universe

Telescopes make things look bigger, too, but they're designed for faraway objects. Curved mirrors gather and focus light, allowing us to study worlds beyond our own.

Without telescopes, we wouldn't understand our place in the universe. We wouldn't know about galaxies, black holes or other cool space stuff. And in a future without telescopes, how will we know if an asteroid is heading towards Earth?!

YOU CHOOSE

Understand the secrets of life on Earth or the secrets of space?

Would you rather travel by steam train ...

Full steam ahead!

The long-distance steam train the Orient Express launched in 1883. It travelled right across Europe, between Paris, France and Istanbul, Turkey. With its luxurious fixtures and fittings, and five-star food on board, a journey on the Orient Express was the experience of a lifetime.

Today, the journey between the two original cities takes about six days. True, you could be there in four hours on a plane, but on the Orient Express, you'd really be travelling like royalty!

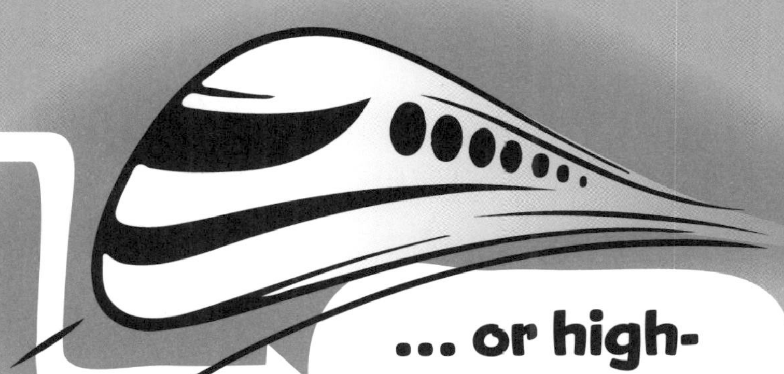

... or high-speed train?

Faster than a speeding bullet ...

The world's first high-speed train was Japan's Shinkansen, which began running in 1964. Back then, the trains reached 210 kilometres per hour. They were nicknamed 'bullet trains' because the front was shaped like a bullet!

Today's bullet trains can reach 320 kilometres per hour. They're still specially designed with an aerodynamic shape – just not the original bullet shape. Now, they have a wedge-shaped nose that cuts through the air with less noise.

YOU CHOOSE

Travel in great style or at great speed?

Would you rather cook your dinner in a kettle ...

Get your food boiling!

Kettles just boil water, right? Wrong! They can also cook food. It's simple – just put a little oil or water in the kettle, then an egg or some rice, perhaps. Yum...

Or perhaps not. The heating element inside the jug switches off automatically when the water's boiling, but it might not work with food. And just think about the bits that flake off the metal element and the limescale that gathers in a kettle. Do you really want those in your dinner?!

... or a dishwasher??

Cook and clean at the same time!

Dishwashers get warm when they're working – up to about 75° Celsius. That's enough heat to cook certain foods, including fish, eggs and vegetables. Just make sure they're sealed tightly in containers.

Dinner from the dishwasher saves energy. It's also surprisingly healthy, because you don't need to add oil or salt. But a dishwasher won't give your food that nice oven-baked crunch!

YOU CHOOSE

Have a boiled brunch or a soggy supper?*

*Don't try this at home!

Would you rather fly around the world in a hot-air balloon ...

Enjoy a bird's-eye view

From a hot-air balloon, you'd get clear view of the landscapes below and the broad horizon. It would be peaceful, too – just the whoosh of the burner sending hot air into the envelope.

However, this isn't a speedy way to travel. A hot-air balloon's top speed is about 16 kilometres per hour, and the world is 40,075 kilometres all the way round. You'd have to make a lot of stops – or have a specially built balloon!

... or a helicopter?

Move in any direction!

At around 260 kilometres per hour, a helicopter definitely travels faster. It would be warmer, too, because it has a closed cabin instead of an open basket. Plus, being able to move up, down and side to side would be fun!

But there are downsides. Rotor blades act like wings, lifting the helicopter into the air. The motion of the blades, together with the engine, make helicopters very noisy. You'd also have to stop and refuel every three or four hours.

YOU CHOOSE

Slow and scenic or fast but loud?

Would you rather challenge a robot to a chess match ...

Man versus machine

Ever since scientists began developing AI in the late 1940s, they imagined a computer that could outsmart the human brain. In the 1990s, IBM rose to this challenge, designing an intelligent computer called Deep Blue. It had one special talent – playing chess.

The computer could analyse around 200 million chess positions per second. But was that enough to beat world champion chess master Garry Kasparov? In 1997, the man and the machine faced off in a series of six games. And Deep Blue emerged the winner!

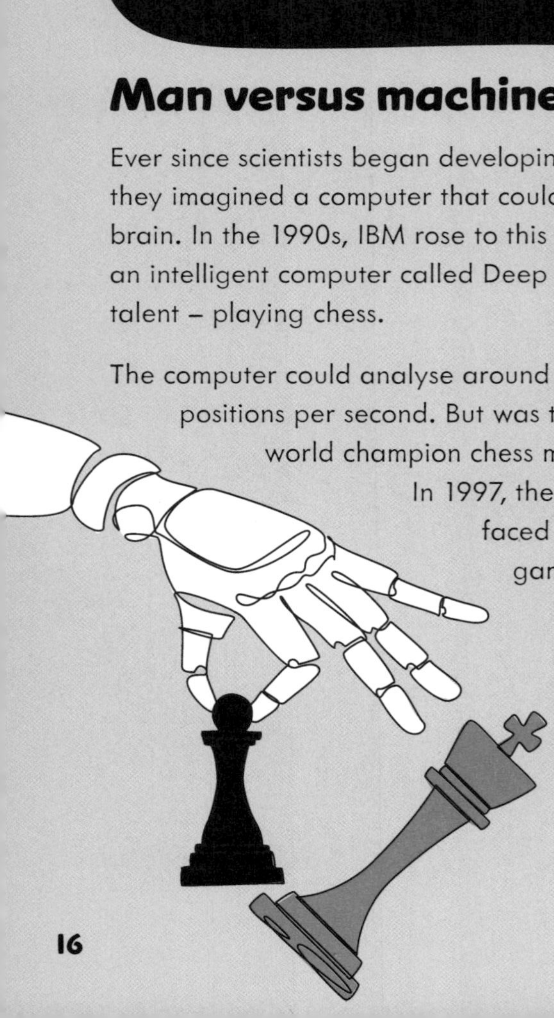

... or a game of Jenga?

Take on a tactical test

In 2019, the world-famous science and technology university MIT developed a robot that's brilliant at Jenga. This block-dismantling game requires both physical and thinking skills.

The robot is patient and careful. It identifies the block it can most safely remove from the tower. Then it edges out the block with the help of a camera, gripper and a 'wrist' that senses how much force is needed.

YOU CHOOSE **Have a brilliant brain or a steady hand?**

Would you rather go into battle in a ball tank ...

Rolling into trouble

Not all tank designs have been winners! Take the German Kugelpanzer, or 'ball tank', which was invented during the Second World War (1939–45). This small, round, armoured vehicle rolled along on raised treads on each side.

It could reach speeds of about 50 kilometres per hour, but it can't have been easy to drive! There was just room inside for a single soldier, who could only see out through a tiny slit in the thin metal ball.

... or a tricycle tank?

You might end up a sitting duck!

The odd Tsar Tank, built during the First World War (1914–18), was nicknamed the 'tricycle tank'. With two huge wheels at the front and a small rotor at the back for balance, it looked like a giant tricycle!

The big wheels were intended to roll over anything in their path. But at 60 tonnes (about the same as 10 elephants!), the tank was too heavy. It kept sinking into the soft ground.

YOU CHOOSE

Face a fight in a rolling tin can or be an easy target stuck in the mud?

Would you rather swallow a nanorobot ...

Take a robotic pill

Nanorobots are *really* tiny robots – smaller than the width of a hair. They don't exist yet, but when they do, they'll be given to patients to swallow, and they'll get straight to work targeting sick cells inside the body.

But how would you really feel about a machine moving around inside you? Could the materials they're made of be harmful? What if the bot malfunctioned? Hopefully, science will solve these problems before letting nanorobots loose!

... or be operated on by a surgical robot?

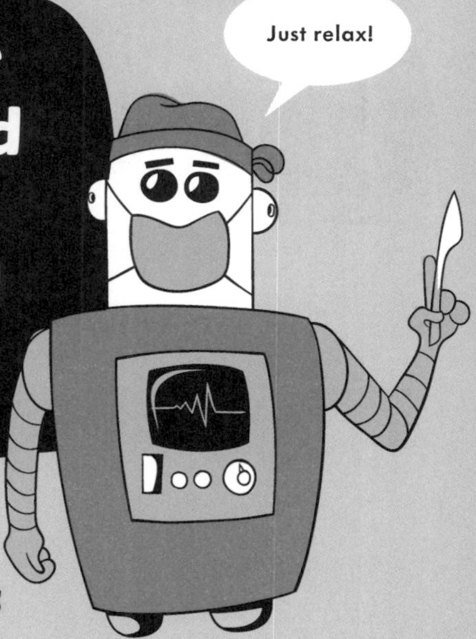

Put your health in a robot's hands

Robot surgeons are already out there in the real world. And they're great at it! Patients recover more quickly from robot surgery. There's less risk of infection and scars are smaller.

But behind the scenes, surgical robots are controlled by real surgeons. And there are still problems. Occasionally, the robots go wrong. Plus, there's a higher risk of things like nerve damage. Maybe human hands are better ...

YOU CHOOSE

Risk a robot breaking down inside your body or while cutting you open?

Would you rather use a flyboard ...

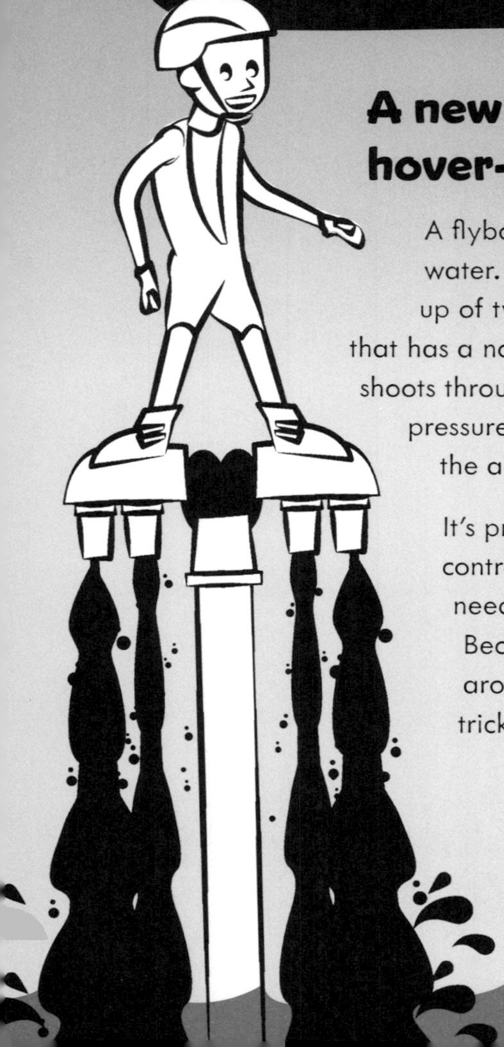

A new type of hover-craft

A flyboard lets you 'fly' above water. The equipment is made up of two boots fixed to a board that has a nozzle on either side. Water shoots through the nozzles to create pressure to propel the rider up in the air.

It's pretty easy to learn how to control a flyboard, and you don't need a lot of special training. Because it's light, you can move around easily and even do tricks like spinning and flipping!

... or a jetpack?

That's one powerful backpack!

Jetpacks are hard to get hold of. And even if you had one, you'd need proper training and safety gear. But once you've strapped on the harness and figured out how the handles work, there'd be no stopping you!

The engine-powered pack means you can go much higher than on a flyboard – nearly 10 metres. And they're fast! You can zoom around at up to 50 kilometres an hour. However, they're not really built for tricks like dives and flips.

YOU CHOOSE

Show off your aerial acrobatics or speed straight to the sky?

Would you rather go supersonic on Concorde ...

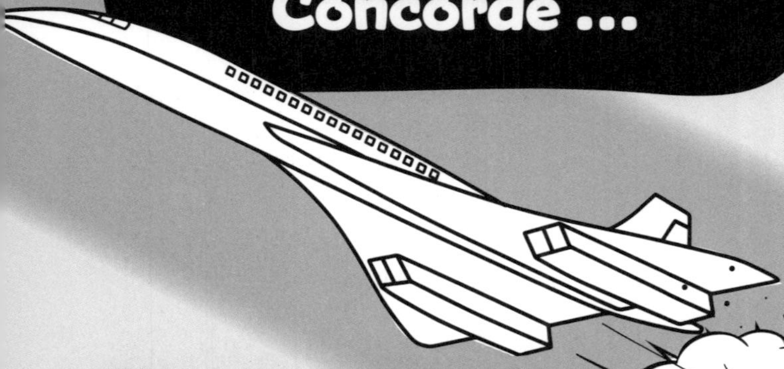

Block your ears for the sonic boom

For a long time, Concorde was the fastest passenger vehicle on the planet. This amazing plane, designed nearly 50 years ago, travelled faster than the speed of sound. It could cross the Atlantic Ocean in just 3.5 hours.

Concorde may have been cool – but it wasn't very comfortable. It had cramped seats and small windows. It was also incredibly noisy, especially when it hit Mach 1 and let out its sonic boom!

... or hypersonic on Falcon 9?

Strap yourself in for a supersonic space ride

Falcon 9 is a launch vehicle designed to put spacecraft and people into Earth orbit. Named after its nine engines, Falcon 9 takes just 70 seconds to reach Mach 1. But it doesn't stop there. This rocket isn't supersonic – it's hypersonic!

Launching into space can be a rough ride. If you can feel the vibrations at Mach 1, just imagine what it's like at 10 times the speed. Plus, there's the danger of sitting on top of all that rocket fuel. Yikes!

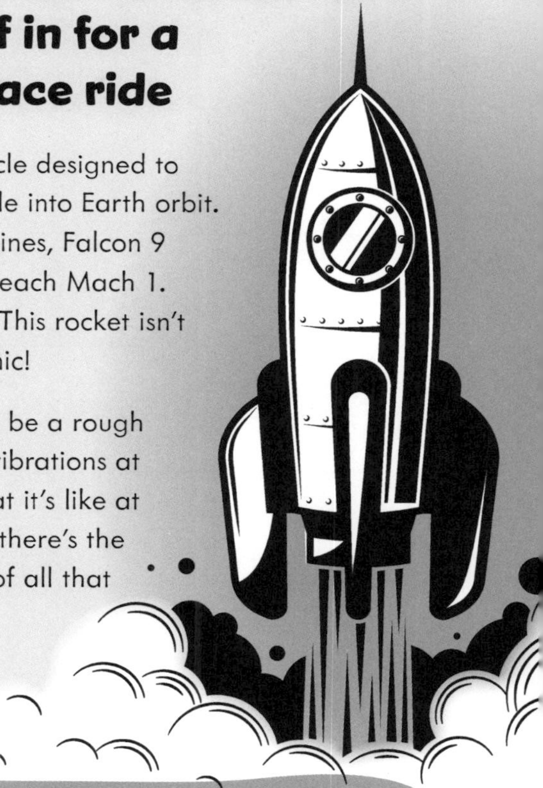

YOU CHOOSE **Classic but cramped or modern but bumpy?**

Would you rather send a secret using Morse code ...

Learn the language of dots and dashes

Before telephones were invented, Morse code was used to send messages over long distances. It used a system of dashes and dots for each letter of the alphabet. The messages were tapped out using a machine called a telegraph, which transmitted electrical signals over wires.

The dots and dashes may seem random, but there is a logic to them! The more common letters have shorter sequences and use more dots than dashes, so the messages are quicker to send.

... or the Enigma code?

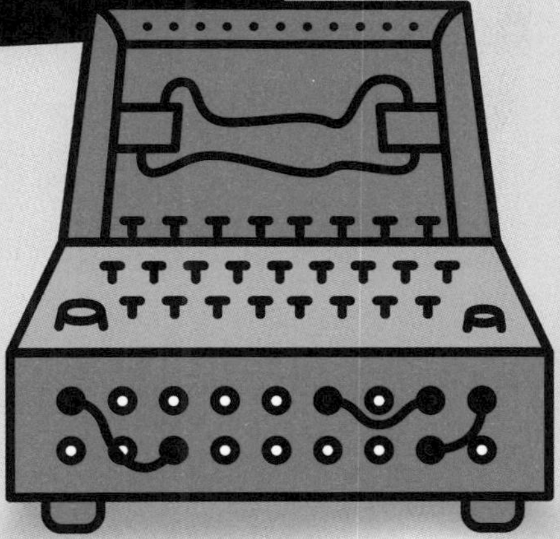

Ciphering and deciphering

During the Second World War, a German engineer invented a cipher machine to send coded messages that would bamboozle the enemy! The Enigma machine looked like a big typewriter. Someone typed in a message, and a system of rotors scrambled the letters. A machine at the other end unscrambled them.

In fact, the British eventually cracked the Enigma code using their own incredible machine – the Bombe. This tested all the possibilities to figure out the 'key' to the messages.

YOU CHOOSE

Quicker to understand but not very secret, or very secret but takes a machine to encode?

Would you rather fall into a cement mixer ...

Keep clear of the concrete!

All sorts of mishaps can happen on a building site! Can you imagine getting stuck in a cement mixer, for example? This type of lorry has a huge turning barrel on its back, which sloshes the cement around and stops it setting.

A tumble inside would be unpleasant. Around the inner wall of the barrel is a spiral blade. This giant corkscrew churns up the cement – and anything else that passes by – before pushing it out of a chute at the other end.

... or fall off a crane?

Watch where you put your feet!

Tower cranes are used to construct tall buildings. They can lift loads as heavy as four elephants, and may be a massive 80 metres high.

The operator has to climb a huge, long ladder inside a structure called the mast to get to the cabin high in the air. The last thing you'd want is to slip on those steps ...

YOU CHOOSE

Get all churned up or dangle from a dizzying height?

Would you rather use only solar panels to generate energy ...

Great if you live in a sunny city!

The sun is a great source of renewable energy. Solar panels are small and quiet. They can be built into houses to give people their own electricity source. Solar energy can also be used in remote places.

But although sunshine is free, solar panels can be expensive to install. They only gather energy when the sun's shining, although most store it in batteries so the lights won't go out at night!

... or use only wind turbines?

Great if you live in a windy wilderness!

Wind turbines are a super-efficient source of energy. You need a steady supply of strong wind to make them work, but at least the wind blows day and night!

However, some people don't like the look of wind farms, and they pose a risk to wildlife such as birds. Wind turbines can also be noisy, and they take a lot of looking after!

YOU CHOOSE

More versatile or more efficient?

Would you rather use a screw ...

An ancient water pump

It wasn't easy being a farmer in ancient Egypt without all our modern machinery. But there was some useful equipment that made life a bit easier. One of these was the Archimedes' screw, which was used for irrigation.

It was made up of a big screw inside a cylinder, with a handle at one end. Put one end of the screw in a river, then turn the handle to raise the water and channel it on to your crops. Hard work – but better than dragging buckets of water around!

... or a shadoof?

Practise your weightlifting

The shadoof was another amazing invention for raising water. A long wooden pole was balanced on a tall upright structure. A bucket hung from a rope at one end of the pole, and a counterweight hung from the other end.

The farmer pulled down on the rope to lower the bucket into the river. When he let go, the counterweight lifted it again. The whole machine could swing around to tip the water out away from the river.

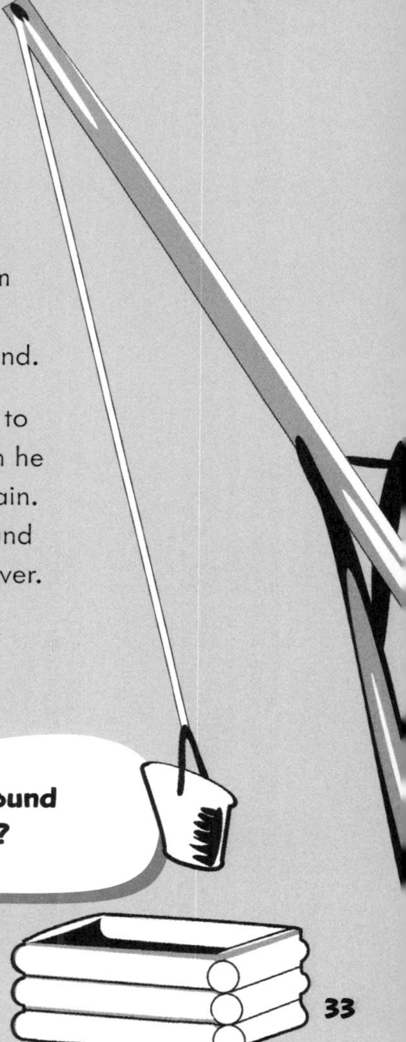

YOU CHOOSE

Go round and round or up and down?

Would you rather have a self-driving car ...

Hands off the wheel!

Can you imagine travelling around in a car with no one in the driving seat? Well, that might be possible in the not-too-distant future.

Autonomous vehicles have cameras and sensors. These mean they can 'see' and sense every object around them, and react accordingly. But although drivers would be able to let the car do most of the work, they'd still have to be ready to take back control in an emergency!

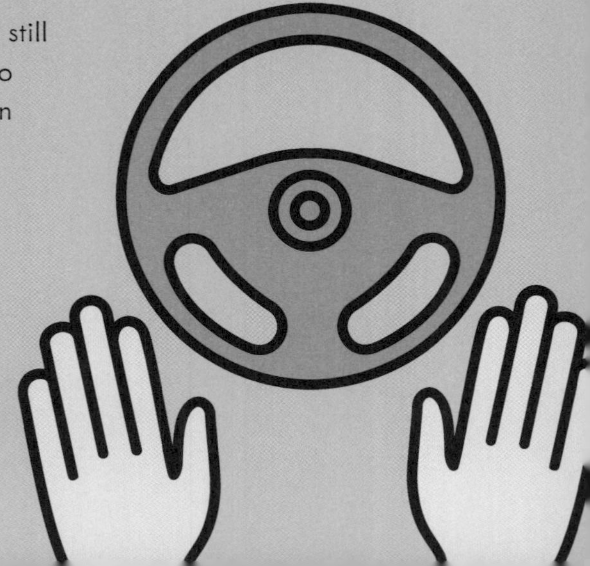

... or a robot chauffeur?

Sit back and relax

IVO stands for Intelligent Vehicle Operator – a robot that can drive cars. It has two struts for legs so it can press the pedals, and a single, three-fingered arm that it uses to steer.

The brilliant thing about an IVO is that it can learn. It can adjust how it drives depending on the car and the road conditions, just like humans do. And it's portable, so you can put your robot driver in any car you like!

YOU CHOOSE **Trust your life to a smart car or an intelligent robot?**

Would you rather have a bionic limb ...

Cyborgs walk among us!

Cybernetic **org**anisms are part human, part machine, which is exactly what people with bionic limbs are. These artificial arms and legs are built for people who have lost theirs through illness or accident.

Some bionic limbs respond to electrical impulses from the muscles. For example, flexing the real muscles in your arm sends a signal to open or close the bionic hand. There are even limbs that respond to electrical signals in the brain and nerves. They can be moved just by thinking about it!

... or a bionic brain?

Fancy a brain upgrade?

Bionic brains aren't real yet, but scientists have created electronic memory cells that work in a similar way to the human brain. These cells can process and store loads of information.

One day, cells like this might be implanted in people who have diseases that affect the brain. They could even be used to recreate brains outside the body for scientists to experiment on!

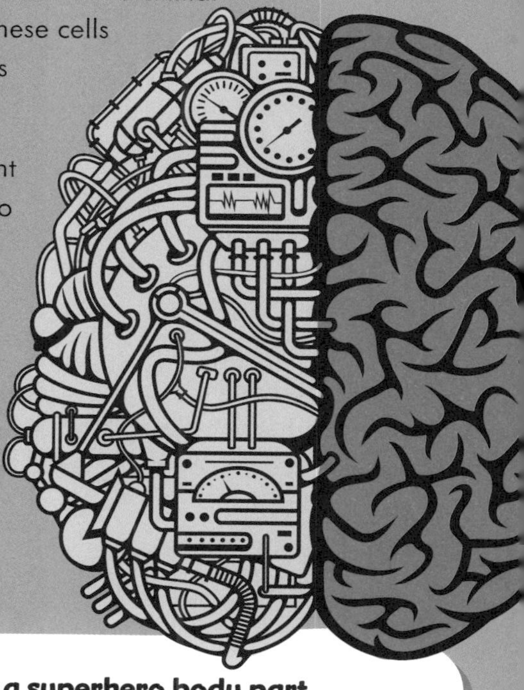

YOU CHOOSE Have a superhero body part or a computer-like brain?

Would you rather live through a GPS blackout ...

Where in the world ... ?

The navigation system (GPS, or global positioning system) in a car or on a phone works thanks to a network of satellites orbiting Earth. They send out signals that your GPS device receives. The device uses the information to work out where you are – within an amazingly accurate 7 metres!

If all the navigation satellites stopped working, traffic would grind to a halt all over the world. But it wouldn't just be drivers going the wrong way! GPS satellites control other vehicles, too, such as planes and ships, which would be lost in the air and at sea.

... or a weather satellite breakdown?

But they said it would be sunny!

Modern life doesn't only rely on GPS satellites. There are other types, too. Geostationary satellites don't move around. Instead, they stay in the same position, taking pictures of Earth every 30 minutes or so.

Those pictures are what experts use to tell you if you need to pick up your sunglasses or an umbrella! Meteorologists identify patterns in the pictures to predict the weather.

YOU CHOOSE

Lose your way or get stuck in a downpour?

Would you rather race a Formula One car ...

Get into pole position

Formula One cars can hit incredible top speeds of 320 kilometres per hour. This is partly due to a very light body, which is made from a single piece of material. The tyres are made of a special rubber that sticks to the track, even when the car is going around sharp bends.

Drivers who race these cars have to stay focused to make split-second decisions while driving at such high speeds. There's no room for mistakes, so it's an adrenaline-fuelled ride!

... or a monster truck?

Power up for a crushing victory!

Powerful monster trucks are modified cars with enormous wheels and tyres that can crush just about anything in their path. They're usually painted in bright colours, and have names like Krimson Krusher and Grave Digger.

They may be big and heavy, but monster trucks are surprisingly agile. They flip around and jump in the air more than twice the height of a double-decker bus. And some of them can hit more than 100 kilometres per hour!

YOU CHOOSE **Take control of great speed or great strength?**

Would you rather ride high on a Maglev ...

A train without an engine

Maglev (magnetic levitation) trains use electromagnets to allow them to 'float' over their tracks. The magnetic field created by the magnets moves the train along. Without any friction between tracks and wheels, Maglevs go very fast!

So, why doesn't every city have a Maglev station? The answer is mainly because the tracks and trains are very expensive. And they'd need to cover long distances to make the most of that super-speed!

... or low on a hyperloop?

Hyper-speed public transport

A high-speed hyperloop would also make use of maglev tech. Instead of an above-ground train, though, passenger pods would zip through a vacuum tube, with nearly all the air sucked out. No air resistance means much higher speeds.

The hyperloop is only an idea at the moment, because it's expensive to install and would take up a lot of land. And be warned! The self-driving pods could move along at more than 1,000 kilometres an hour. That's certain to make you travel sick. Bleurgh!

YOU CHOOSE

A long wait for the next train or vomit in a vacuum?

Would you rather have a robot pet ...

Not-so-furry friends

These days, you don't have to commit to looking after a real cat or dog – just get a robot pet! You wouldn't have to take it for a walk if you were tired. There would be no expensive vet's fees. And you wouldn't have to pick up its poop!

But robot animals aren't very furry or cuddly – and that's half the fun of a pet. You'd also know deep down that the wagging tail or contented purr was just a computer program.

... or a robot friend?

Almost human

Humanoid robots are getting more and more realistic. Take Sophia, for example – the world's first 'robot citizen'. She's a cybernetic human, designed to look and behave in an eerily lifelike way.

Sophia is programmed with lots of realistic facial expressions. She also uses AI to recognise and respond to other people's expressions. She can copy emotions to interact with people in a sympathetic way. Sophia is also very clever and can happily hold an intelligent conversation!

YOU CHOOSE

A no-hassle animal or a friend who'll focus on you?

Would you rather be squashed by a vending machine ...

Don't get greedy!

It's a strange fact that more people die from vending machines falling on them than they do from shark attacks. To avoid a 450-kilogram box of metal toppling over on you, treat it with respect!

Injuries occur when people get greedy or impatient. They tip the machine up to try and get something out without paying. Or the mechanism that releases the chocolate gets stuck, and people rock the machine to jiggle it out. Both are a big mistake!

... or get your arm stuck in a claw machine?

Cheaters get caught!

You've probably seen claw machines at amusement arcades. A joystick controls a claw that tries to grab a prize inside a big glass or plastic box. But the claws are never as strong as they look, so it's hard to keep hold of that cuddly toy!

Annoyed players sometimes stick their arm in the prize-drop slot to try and win another way. And they often get stuck! Sometimes children manage to climb all the way into the enclosure – then can't get out again!

YOU CHOOSE

Get caught stealing or get stuck cheating?

Would you rather live without computers ...

Everything would be a lot more effort!

The first digital computer (ENIAC) was built in 1946. Since then, we've come to depend on computers so much that, without them, the world really would come to a standstill.

The internet wouldn't exist. You'd have to find information in books or other printed material. There would be no online shopping or food delivery services. You could also say goodbye to video streaming and social media. Is that the kind of world you want to live in?

... or mobile phones?

Is there anyone there?

It's strange to imagine, but 25 years ago hardly anyone had a mobile phone. Today, almost everyone has one of these mini machines that keep us in contact everywhere we go.

Without a mobile, you'd have to rely on landline telephones. There would be no texting, social-media messaging or video calls. If you needed to make a call when you were out and about, you'd have to use a payphone. And you'd need to find another way to listen to music or play games on the go!

YOU CHOOSE

A life without instant information or without constant contact?

Would you rather ride a penny farthing ...

How the Victorians cycled

This odd-looking bicycle was popular in the late 1800s. Because it had one small wheel at the back and one huge one at the front, it was named after the smallest and largest coins of the day – the penny and the farthing.

These bikes were high, so riders got a good view of the road. The simple design also made the bike easy to maintain. But their height meant that it was a painful fall if you hit a bump and tumbled off. You also needed a stool to get on and off!

... or a unicycle??

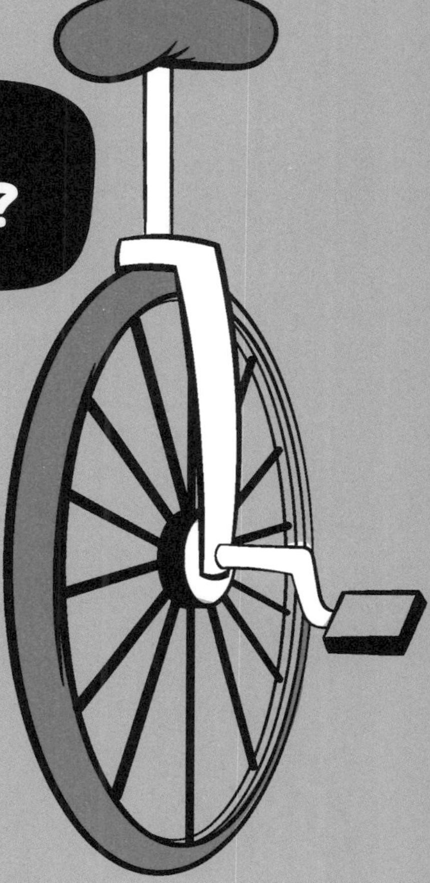

A balancing act

One-wheeled unicycles are as tricky to mount as a penny farthing. And once you're on, you need excellent balance to stay upright! It takes a lot of practice and patience.

But if you can get to grips with it, a unicycle is a great way to get around, as they can nip around obstacles quickly and easily. You could even learn some tricks and stunts to impress your friends!

YOU CHOOSE **Amaze your friends with a two-wheeled curiosity or some single-wheeled stunts?**

Would you rather go on the world's highest drop tower ride ...

Yes, people do this for fun!

Drop tower rides have a tall central structure, with rails up the side. Brave passengers are strapped into gondolas that are lifted to the top of the tower. They're then released into free fall! Luckily, brakes kick in near the bottom of the ride to stop it crashing to the ground.

The Zumanjaro ride in New Jersey, USA, is the highest in the world. It drops people a terrifying 126 metres, reaching a speed of 140 kilometres an hour!

... or the world's fastest roller coaster?

Feel the force

The Formula Rossa in Abu Dhabi isn't for nervous types! During the ride, the roller coaster climbs to a height of 53 metres. But that's not the bit that takes your breath away ...

It reaches 240 kilometres an hour in less than 5 seconds. That creates a g-force of 4.8. For comparison, astronauts usually experience a g-force of between 3 and 4 during launch!

YOU CHOOSE **Face a heart-thumping free fall or a body-squashing g-force?**

Would you rather ride on a funicular railway ...

Heading straight up!

Funicular railways are used to carry people and goods up super-steep slopes, like the side of cliffs and mountains. Two trains run on wheels on steel tracks like an ordinary railway, but they're pulled along by a cable.

The cable loops around a pulley at the top. As one train goes down the cliff, the other is pulled up, like counterweights. Some funicular railways climb slopes of around 45 degrees, which makes for a nail-biting ride!

... or in a cable car?

Hope you're not afraid of heights!

How do you fancy a high-wire ride in a cable car? The cars are basically metal and glass boxes that hang from an arm gripping a steel cable high above the ground. Passengers get on and off at turning stations at either end of the system.

It seems a safe enough way to travel, but cable cars seem designed to make you nervous. Some cars have glass floors so you can see right down from the dizzying heights!

YOU CHOOSE

Get vertigo from a vertical climb or from dangling in mid-air?

Would you rather stay in a robotel ...

A hotel with staff problems!

When it opened in 2015, the Henn na Hotel in Japan was mostly staffed by robots, from a robot receptionist to automated trolley porters. The idea was to make the hotel super-efficient – but it didn't work out that way!

The robots broke down quite often, so they couldn't do their jobs. As a result, more than half of them were 'fired'! Guests also found the robots annoying. They were hard to understand, and the humans ended up doing a lot things for themselves!

... or eat at a robo restaurant?

Dinner is served

Robot restaurants use machines for some jobs that humans usually do, including robot waiters and chefs. Some robot restaurants have done well, thanks to their machine staff offering good food and good service.

But others have been a disaster. Sometimes, robot chefs take a really long time to make food! Plus, machines aren't always cheaper than people. And of course, there's the chance that a machine will malfunction in the middle of making your meal!

YOU CHOOSE

A bed-making machine that keeps breaking down or a long wait for bad food?

Would you feel safer piloting a stealth plane ...

Make a sneak attack from the sky

Stealth fighters are specially designed so that they can't be identified by enemy radar. They're very speedy, too, which means they can do their job and escape very quickly.

But flying a stealth plane is still a risky job. The planes can't be detected by radar, but they can be picked up by infrared sensors. And they can only carry a limited amount of fuel and weapons. So, if a mission doesn't go as planned, you could be in great danger.

... or a stealth ship?

Almost undetectable ...

Most stealth ships are very hard to detect by radar, sonar or infrared sensors. They're also very strong and can usually keep operating even under enemy fire.

But stealth ships are bigger than other ships, which means that they can't manoeuvre as easily. If they come under fire from smaller boats or aircraft, they might struggle to dodge the attack. They're slower than other types of ship, too, which isn't great for a quick getaway.

YOU CHOOSE **Make a quicker escape or be harder to spot in the first place?**

Would you rather live in a smart home ...

Take control of your whole life!

Imagine coming home from school and the lights automatically turn on for you, your favourite music is playing, and your bedroom is set to the perfect temperature. Welcome to your smart home!

You could control your smart devices with simple voice instructions. Want to turn on the TV, get the microwave going or close the curtains? Just ask! Plus, smart security systems make smart homes very safe.

... or drive on a smart road?

Get there quicker and safer

Smart roads use technology to improve safety, the flow of traffic, and the whole driving experience! Sensors and cameras monitor the amount of vehicles. They then adjust signs and traffic signals to help things along.

By identifying dangerous conditions, such as ice or obstacles on the road, smart roads can make driving safer. And they can be environmentally friendly, too. Some smart roads gather renewable energy to power lights and traffic signals!

YOU CHOOSE **Enjoy an easy home life or a safer, more efficient journey?**

Would you rather see your skeleton ...

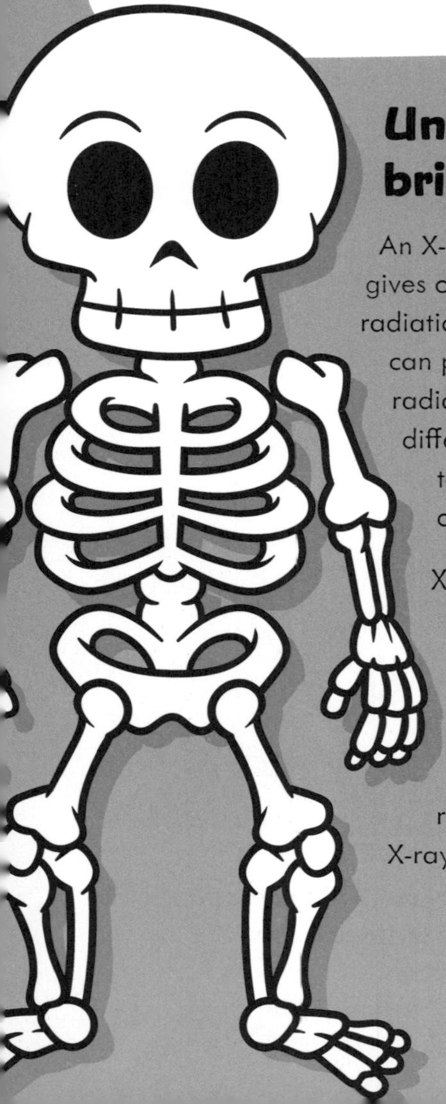

Understand your brilliant bones

An X-ray machine is a device that gives out a small amount of ionising radiation – a type of energy that can pass through your body. The radiation interacts differently with different parts of your body. It's used to create a special type of film to create an image of your insides.

X-rays are quick and painless. They're used by doctors to diagnose everything from broken bones to pneumonia to dental problems! However, any exposure to radiation can be risky, so you wouldn't want to be X-rayed too often!

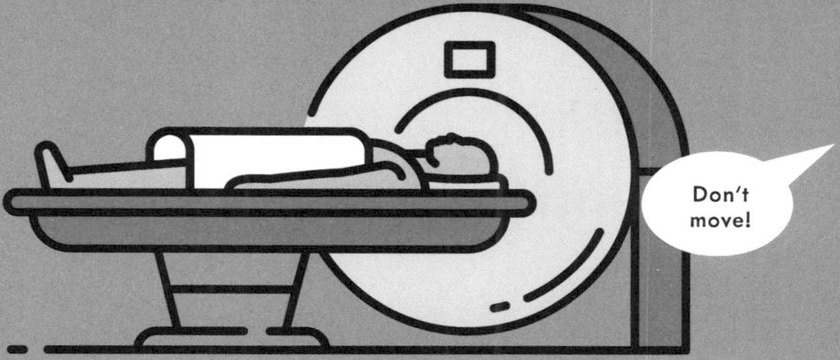

... or your organs?

Get a glimpse of your amazing organs

MRI stands for 'magnetic resonance imaging'. The machine uses a strong magnetic field and radio waves to create images of your tissues and organs.

The procedure is painless, and it can identify a whole range of problems. But it can take a while to complete, and patients have to lie completely still all that time. It can also be very noisy, as the machine makes banging or knocking sounds while it takes its pictures.

YOU CHOOSE

Check that your bones are strong or your organs are working properly?

Would you rather be trapped in a lift with someone who's farted ...

Hold your nose!

Lifts are designed to be pretty airtight. That's because the air pressure can change when the lift goes up or down, which can make passengers' ears pop – like on a plane. So, lift doors create a tight seal when they close. That can make it pretty unpleasant if you're stuck with someone who's let off a bad smell!

Fortunately, lifts usually have a ventilation system. This helps air to circulate inside and removes stale air. At least that would save you from standing in a stink for too long!

... or in a reduced-gravity machine with someone's who's been sick?

Be prepared to lose your lunch

A reduced-gravity machine is a special aeroplane used to prepare astronauts for zero-gravity conditions. And it's not nicknamed the Vomit Comet for nothing!

Once the astronauts are inside, the plane climbs steeply upwards. As it drops into a sudden free fall, it can cause severe motion sickness. And if someone's sick, it's not easy to avoid the chunks when you're floating in zero gravity!

YOU CHOOSE

Try not to breathe in or play 'dodge the vomit'?

Would you rather travel by Segway ...

It's a bit of a balancing act!

The Segway was the world's first 'self-balancing human transporter'! You just have to lean backwards or forwards slightly to move the machine in that direction. Turn the handlebar to go round a corner.

With their big wheels, Segways are a surprisingly smooth ride – as long as you're on a road or pavement. They're not designed for rough surfaces. Or big hills, for that matter! Segways are also battery-powered, so you can't travel far without having to stop and charge up.

... or scooter?

It might be a bumpy ride

Light, battery-powered scooters are a great way to get around. You just use the touchscreen on the scooter to tell it what to do!

Being lighter makes a scooter less stable than a Segway, and they need more skill to operate. They also have a speed limit and aren't great for long distances. And they can be an uncomfortable ride, even if you're just going a little way!

YOU CHOOSE

Balance your way over bumps or go for a slow, unstable scoot?

Would you rather listen to music on a gramophone ...

Get the music spinning!

A hundred years ago, people listened to music on machines called gramophones. Songs were recorded on to vinyl disks, which spun on the gramophone while a needle ran round the grooves and played the songs.

Gramophones produced a beautiful sound – some say it was even better than digital! But the records could easily be scratched, which made the songs 'jump' and reduced the sound quality. They could also only contain around 30 minutes of music on each side.

... or a cassette player?

The start of music on the move

Cassette players became popular in the 1970s. Music was recorded on to a tape wound inside a plastic cartridge (cassette). A cassette player 'read' the tape and played the music through speakers or headphones.

Cassettes could hold more music than records – up to 90 minutes per side. But the tape would eventually wear down, reducing sound quality. If you wanted to skip a song, you had to fast forward or rewind, which was time-consuming and annoying!

YOU CHOOSE

Put up with scratchy songs or tangled tape?

Would you rather have a robotic vacuum cleaner ...

Hassle-free housework

Robotic vacuum cleaners use sensors to find their way around and stop them from bumping into things. Downward-pointing sensors stop the cleaner falling down stairs, for example. Rotating brushes sweep up dirt as they go.

These helpful machines have changed how people do housework. They can save time and are very efficient. But the fact is, they're not as good as traditional vacuums with a human in control. And they need to be emptied more often!

... or a robotic lawnmower?

No-effort gardening

Robotic lawnmowers potter about the garden keeping the grass trim. A guide wire around the edge keeps them in the right place. They run on electric motors, and some can even find their way back to their charging stations when they need a boost.

The problem is, they're only really any good for small or medium-sized gardens. They'd be overwhelmed by a lot of lawn. They can also pose a risk to pets, so don't let the guinea pig out when the lawnmower is on the loose!

YOU CHOOSE

A clean house or a tidy lawn?

Would you rather tear down a building using a bulldozer ...

Squashing small structures

If you want to demolish a smallish structure, get behind the wheel of a bulldozer. It has a large metal blade on the front, which can push something until it collapses. It can also be fitted with a 'crusher' that can smash through bricks.

The best thing about a bulldozer is that it's not only designed for knocking things down, but also for clearing things up! The blade can be used to scrape up the mess, and tilted to carry it to a dump site.

... or a wrecking ball?

Aim carefully!

A wrecking ball is a heavy steel ball attached to a chain or cable, which is hung from a crane. The crane's arm swings the huge ball against a building to bring it crashing down.

The biggest balls can weigh more than 5,000 kilograms, so if you're going to crush a building with a wrecking ball, you need to plan it carefully. After all, you don't want to hit the wrong building – or anyone standing nearby!

YOU CHOOSE **Crush buildings beneath your wheels or with a terrifying ball?**

Would you rather work on the world's biggest computer ...

Air defence without human error

The largest computer ever built was the Semi-Automatic Ground Environment (SAGE). Completed in 1963, this incredible machine was actually 56 computers all working together. In total, it covered an area of 1,800 square metres – that's about seven tennis courts!

SAGE was actually an air-defence system designed to detect and stop enemy aircraft. The computers were connected to radar stations around the USA. If a threat was detected, they would launch fighter jets to intercept the enemy plane.

... or the world's smallest?

Small but powerful

The Micro Mote might be the smallest computer ever created. It's so tiny – just 1 millimetre wide – that it can fit on the tip of your finger! But in that teeny structure is a computer chip, memory and sensors that can measure things like temperature and pressure.

This micro machine could be really useful. For example, it could be included in medical implants to help doctors monitor someone's health. It could also be used to keep track of things like air and water quality, to help save our planet.

YOU CHOOSE

Aching legs from running around the largest computer, or aching eyes from squinting at the tiniest?

Would you rather race across the desert ...

Navigate the sand dunes

Sand is tricky to travel across. The tiny grains make for a slippery surface that vehicles can easily sink into. That's where dune buggies come in. These small, light cars have big wheels and wide tyres, which stop them getting stuck.

They're a lot of fun to drive, but their open tops mean there are certainly safety issues. Firstly, you're exposed to the blazing desert sun. Also, because they don't have seat belts, if the buggy overturns you'll probably come a cropper!

... or speed across snow?

An exhilarating ride

Snowmobiles have tracks and skis instead of wheels, so they glide across snow and ice. But they have an engine and a clutch like normal road cars and bikes. Most snowmobiles can power along at around 100 kilometres an hour, so you'll need proper safety gear, including goggles!

Sadly, snowmobiles aren't great for the environment. They can cause air pollution. And because they can be quite noisy, they can scare wildlife that lives in snowy regions.

YOU CHOOSE

Risk severe sunburn or snow blindness?

Would you rather spend time in space with a robonaut ...

The future of space medicine?

Robonauts are humanoid robots designed to help humans in space. They can do everything from assisting astronauts with equipment repairs to carrying out scientific experiments.

NASA's Robonaut 2 (R2) has been working on the International Space Station since 2011. Most astronauts aren't trained medical doctors, so NASA hopes that one day the robonaut will be advanced enough to work as a kind of space doctor, controlled by real doctors back on Earth.

... or a Mars rover?

Scientists on the red planet

The Mars rovers are a series of robotic vehicles that have been sent to the red planet to explore its surface and do scientific research. The first rover arrived in 1996, and others followed, including the Opportunity, Curiosity and Perseverance.

The rovers are equipped with special scientific instruments, such as cameras and drills. These allow them to study the Martian landscape and collect information about its geology, atmosphere and climate.

YOU CHOOSE **A helpful humanoid assistant or a tireless space scientist?**

Would you rather set a book on an old-fashioned printing press ...

One typo and you have to reset a whole page

On early printing presses, all the letters and punctuation marks were cast on metal blocks. They were then put together in a wooden frame to form a page of words. Ink was rolled over the letters, then a piece of paper was pressed on top.

Putting together all the words was very time-consuming, and it needed skilled workers to operate the press. But for the first time, a machine made it possible to create lots of copies of a book quickly and cheaply.

... or type one on a manual typewriter?

No spell check!

Typewriters had a set of keys with letters and punctuation marks, a bit like a computer keyboard. Pressing a key made a small metal 'type' with the letter on it swing upwards and hit an inked ribbon against a piece of paper.

It took some effort to hit those metal keys. And at the end of every line, you had to pull a lever to send the carriage that held the paper back to the start. All that clattering made the typewriter a noisy machine!

YOU CHOOSE

Build every word with blocks or get arm ache hitting every letter?

Would you rather work alongside a cobot ...

Cool colleagues!

Collaborative robots, or 'cobots', are specially designed to work alongside people in a shared workspace, such as a factory or a warehouse. Unlike traditional robots, which are often big and dangerous, cobots are smaller and safer.

Cobots have sensors that can tell if humans are nearby, and they can move around to avoid hurting anyone. And they're excellent multitaskers! They can be programmed to perform a variety of jobs, from selecting and packing items to fixing other machines.

... or a robotic arm?

An arm-y of workers!

Robotic arms are a common sight in factories. They're used when people need a robot to stay in one place and do a single repetitive job. The job can be very complicated – but a robotic arm never makes a mistake!

The robots are made up of several parts with different joints, which means they can move in different ways. They can reach for objects and do different things with them. They can even be programmed to work on their own, without humans telling them what to do!

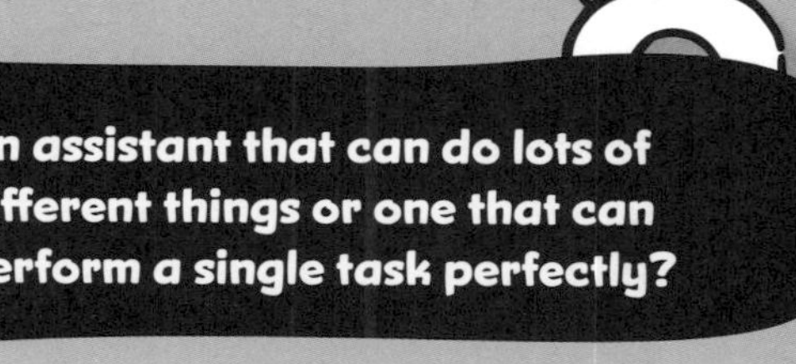

YOU CHOOSE

An assistant that can do lots of different things or one that can perform a single task perfectly?

Would you rather be dumped on by a dump truck ...

Stand back!

Dump trucks are very useful for moving heavy loads from one place to another. To empty the truck, the driver pulls a lever to activate the rams. These are metal rods that move in and out of cylinders to push up the back body to let the load slide out.

It's wise to stand well clear of a truck dumping its load. Bigger vehicles can hold more than 12 tonnes of rocks and other rough stuff. That's the same as being squashed by a double-decker bus!

... or be chased by an excavator?

You won't dig this!

Run for your life if you see an excavator heading your way! These machines can be used for all sorts of tasks, from drilling to digging. The arm can move up and down, and the cabin can swivel right the way round.

An excavator's bucket can scoop up more than 300 wheelbarrows of dirt in one go – and you don't want to get caught up in that! But with speeds up to 30 kilometres per hour, you'd have trouble outrunning one ...

YOU CHOOSE **Squashed by rocks or scooped up with the soil?**

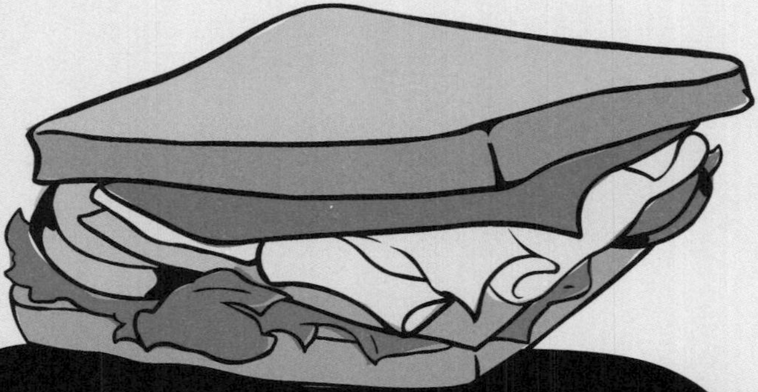

Would you rather eat a sandwich that's been printed ...

A sandwich of many layers

3D printing creates three-dimensional objects by adding layers of material on top of each other. Usually, that material is plastic, but now you can also use edible materials, such as chocolate, dough or pureed fruit, to make a 3D printed meal!

Simply choose your food and load it into a 3D printer. The printer heats it, so it can be squashed through a nozzle, then it follows computer instructions to create the food layer by layer. Of course, this takes some time!

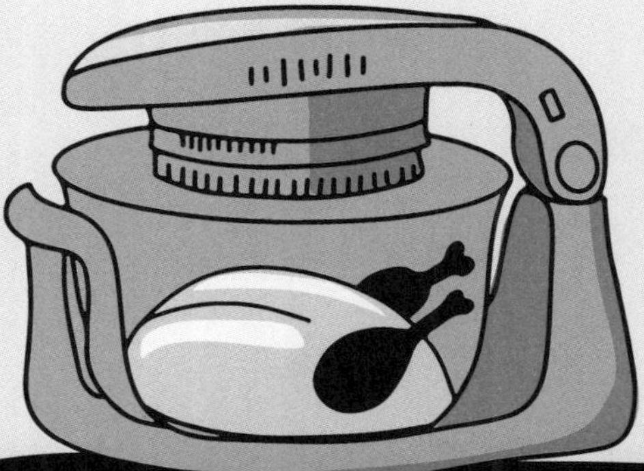

... or air fried?

Beware of the burn!

From ovens to toasters to blenders, machines have always been used to prepare and cook food. One popular new kitchen gadget is the air fryer, which is a healthy option because you don't need any oil.

You could use an air fryer to toast a sandwich, but be careful! It's easy to burn food in one of these fryers. Food also dries out quite easily because it's being cooked with hot air.

YOU CHOOSE **Hang around while your sandwich is built or risk over-toasting?**

Would you rather get stuck in mud ...

Saved from a sticky end!

Rescue workers use all sorts of machines. Imagine you get stuck in mud or quicksand and start to sink, for example. In this case, your rescuers might help you out with a mud lance.

They'd stick the pointy end of this long pole into the mud or sand to create a hole. Then they'd drop in a rope and pull you out. Water or air can also be shot through the hollow pole to stop the mud from sucking you down.

... or buried in snow?

Send out a life-saving signal

If you're venturing into avalanche territory, it's best to be prepared. These huge rivers of snow can come crashing down at up to 320 kilometres an hour!

If you're trapped in snow, a transceiver, or avalanche beacon could save your life. The small device can be attached to a jacket – and when activated, it gives off an electronic signal. Rescue teams carry receivers that can pick up the signal to find out where you are.

YOU CHOOSE

Trust your life to a long pole or a beeping machine?

Would you rather have a teleporter ...

Explore the world – and beyond

Of course, no one has invented a teleportation machine yet. If they did, you'd be able to transport yourself to another place in milliseconds, as long as you didn't mind your body being broken down to atoms and then put back together!

You'd never have to wait for the bus again. Want to see the summit of Everest or stand at the South Pole? With a teleporter, you could get there without a hard climb or a long trek. You could even transport yourself to another planet!

... or a time machine?

Explore the past and future

No one has invented a time machine yet, either. Some scientists say it might be possible, and others say never in a million years. But where would you go and what would you do if you did have a time machine?

You could see how the pyramids were built or taste the food at a medieval banquet. You could save lives by warning people about disasters that haven't yet happened. And how great would it be to visit the future and see what marvellous machines people will one day come up with?

YOU CHOOSE **Instantly appear in another time or another place?**

GLOSSARY

adrenaline a substance in the body that makes your heart beat faster, which is released when you are excited or nervous

aerodynamic having a shape that moves efficiently through air

agile able to move quickly and easily

AI artificial intelligence – the ability of computers and robots to perform intelligent tasks

autonomous describing something that is able to make decisions for itself

avalanche a mass of snow, ice and rocks that falls quickly down the side of a mountain

bionic describing artificial body parts made of electronic and mechanical parts

carbon footprint a measure of the amount of carbon dioxide released by different activities

clutch the pedal in a car that connects the gears

counterweight a weight designed to balance out another weight

cybernetic describing things related to communication in machines

element part of an electric device containing a wire that the electric current passes through to create heat

envelope the 'balloon' part of a hot-air balloon, made of a material that holds the gas

friction the force that makes it difficult for one object to slide along the surface of another

g-force a force that pushes you backwards when moving forwards very quickly

gondola the hanging part of a machine, in a fairground ride, for example, that people sit in or on

horizon the line where the surface of Earth appears to meet the sky

humanoid a machine that looks and behaves like a real person

hypersonic describing things that are more than five times faster than the speed of sound

implant to insert something in someone's body

infrared a type of invisible light that can be used to detect things in the dark

intercept to stop or catch something or someone before they reach a particular place

irrigation adding water to fields to help crops grow

landline a telephone that works using cables that have been laid across land

lens a piece of glass polished and shaped in a curve to focus or send out light

levitation the process of rising and floating

limescale a hard white substance left behind by water that builds up inside pipes, kettles, etc.

luxurious very comfortable and expensive

Mach a measurement of a vehicle's speed in relation to the speed of sound; for example, Mach 1 is the speed of sound and Mach 2 is twice the speed of sound

magnetic field an area around something magnetic that contains a force that attracts objects towards it

malfunction when a machine stops working the way it's supposed to

manoeuvre to move with skill and care

mechanism a part of a machine or a set of parts that all work together

mishap an unlucky event or accident

modified describing something that has been slightly changed to make it work in a particular way

navigational describing things that help you find your way

nozzle a round spout at the end of a pipe or hose, used to control a jet of liquid

orbit to move around a body in space in a curved path

payphone a public telephone that works by putting coins in

pole position the best starting position in a race, in the front

portable small and light enough to be carried around

porter someone whose job is to carry things, especially people's bags at a hotel or railway station

porthole a small, often round window in the side of a ship

pulley a wheel with a groove around its rim that a cable or cord passes around, used to lift heavy weights

quasar a huge space body a bit like a star, which contains lots of energy

radar a system that uses radio waves to find out where things are when they cannot be seen

renewable energy energy from a source that will not run out, such as the sun or wind

rotor part of a machine that spins, often with blades attached to lift the machine

scenic showing beautiful natural surroundings

sequence a series of related things or events

sonar equipment that uses sound waves to detect things in water

sonic boom a loud sound caused by the shock waves created when a vehicle overtakes the speed of sound

statistic facts involving numbers, such as information about the number of times something happens

stealth describing movement that is quiet and careful, in order to not be seen or heard

strut a strong rod or bar, usually made of metal, that helps to hold something together

submariner a sailor who works on a submarine

submerged to be below the surface of water

supernova the explosion of a star at the end of its life cycle

supersonic describing things that are faster than the speed of sound

telegraph a system of communication sending pulses of electricity through wires

tread the raised part of a vehicle, usually on the tyre, that grips the road

turbine a machine that uses moving air or water to turn a wheel and produce power

vacuum a space that all the air has been removed from

ventilation a system that moves fresh air around a closed space

vertigo a feeling of spinning when looking down from high up

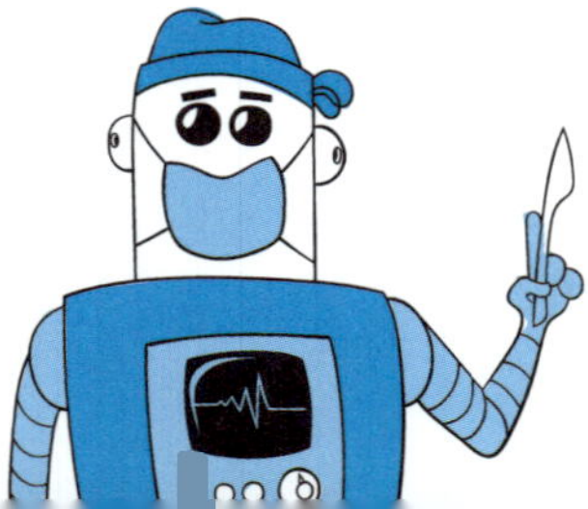

Further Information

Books

Working with Machines **(Kid Engineer)**
by Sonya Newland, Wayland, 2021

Machines in Space/Machines on Land/Machines at Sea **(Engineering Power!)**
by Kay Barnham, Wayland, 2020

Machines and Motors **(Infographic: How It Works)**
by Jon Richards and Ed Simkins, Wayland, 2018

Websites

www.livescience.com/49106-simple-machines.html
Learn about the six simple machines and how they work.

kids.britannica.com/kids/article/car/352807
Read about how cars are designed and built, and how the different parts of these machines work.

www.turito.com/blog/robotics/interesting-facts-about-robot-for-kids
Find out more about robots and robotics – their history and how they're used in the world today.

INDEX

Other titles in the You Choose series

978-1-5263-2423-8

978-1-5263-2433-7

978-1-5263-2430-6

978-1-5263-2434-4

978-1-5263-2429-0

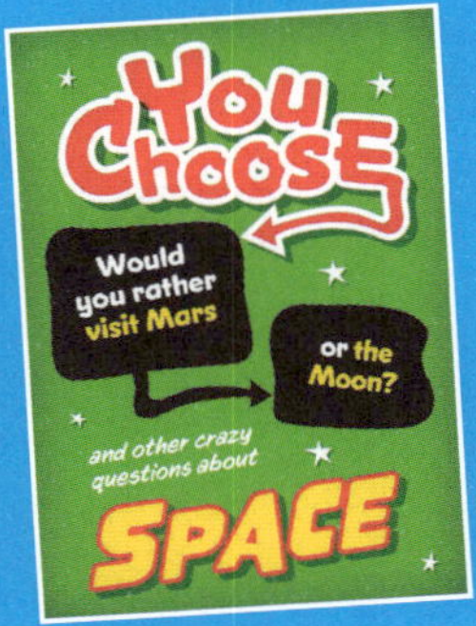

978-1-5263-2427-6

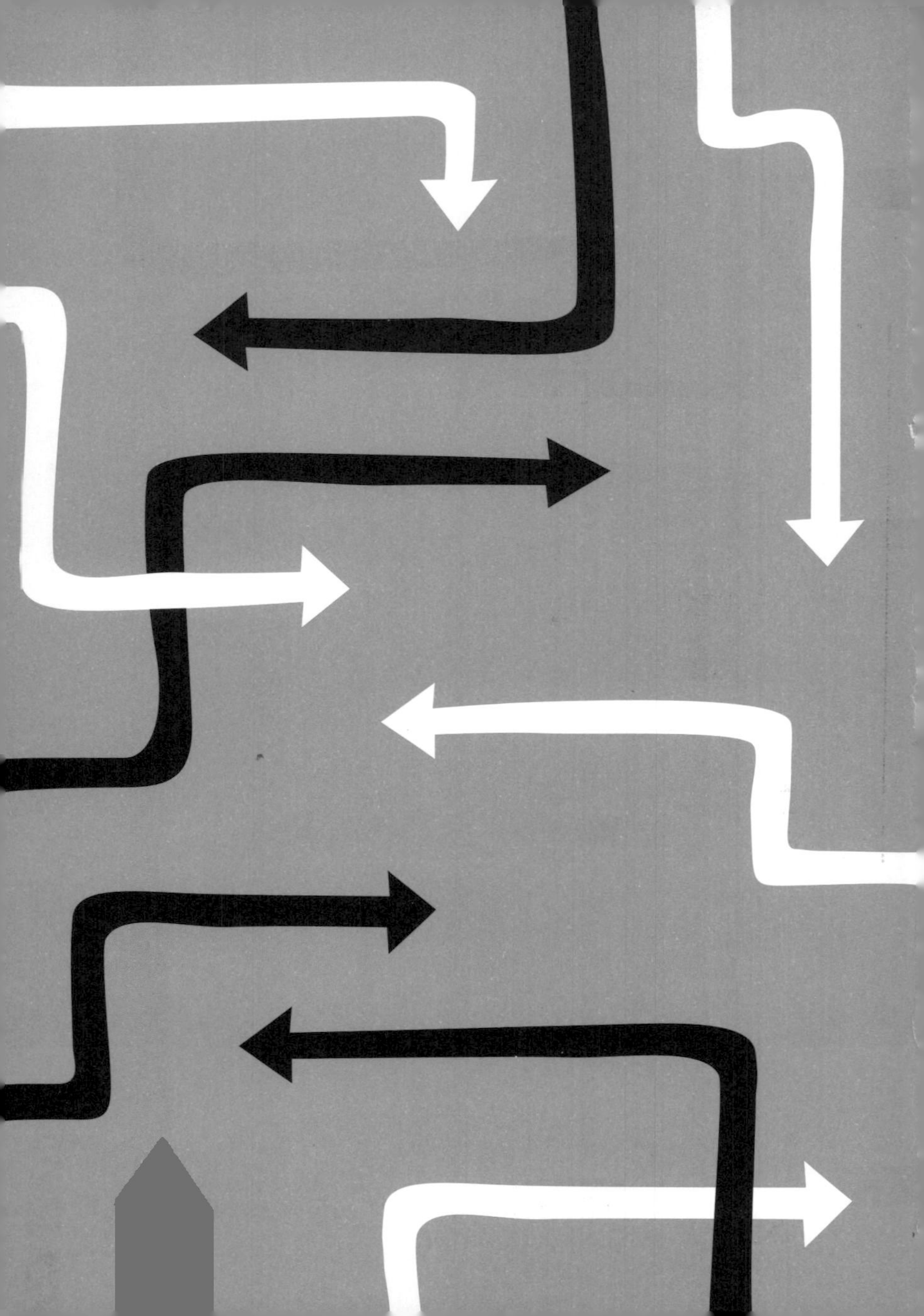